Problem Solver II™

Student Workbook

Judy Goodnow
Shirley Hoogeboom

Acknowledgments

We wish to thank Marj Santos for reviewing the manuscript and guiding the classroom testing.

Judy Goodnow has authored and coauthored over 100 books and software programs for mathematics and problem solving. She has taught children from kindergarten through sixth grade. She holds a bachelor of arts degree from Wellesley College, a master's degree from Stanford University, and a California Teaching Credential from San Jose State University.

Shirley Hoogeboom has authored and coauthored over 100 books for mathematics and language arts. She has been a classroom teacher, and has conducted workshops for teachers on problem solving and on using math manipulatives. She holds a bachelor of arts degree in Education from Calvin College, where she earned a Michigan Teaching Credential. She completed further studies at California State University, Hayward, where she earned a California Teaching Credential.

 **Wright Group**

Problem Solver II: Integrating Problem Solving with Your Math Curriculum, Student Workbook, Grade 3
Copyright ©2004 Wright Group/McGraw-Hill
Text by Judy Goodnow and Shirley Hoogeboom
Illustrations by John Haslam
Design and Production by O'Connor Design

Problem Solver™ is a trademark of the McGraw-Hill Companies, Inc.

Published by Wright Group/McGraw-Hill, a division of the McGraw-Hill Companies, Inc. All rights reserved. No part of this publication may be reproduced or distributed in any form or by any means, or stored in a database or retrieval system, without the prior written consent of Wright Group/McGraw-Hill, including, but not limited to, network or other electronic storage or transmission, or broadcast for distance learning.

Wright Group/McGraw-Hill
One Prudential Plaza
Chicago, Illinois 60601
www.WrightGroup.com
Customer Service: 800-624-0822

Printed in the United States of America.

10 9 8 7 6 5 4 3

ISBN: 0-322-08814-3

The McGraw-Hill Companies

Problem-Solving Strategies

 **Act Out or Use Objects**

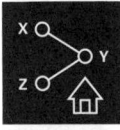 **Use or Make a Picture or Diagram**

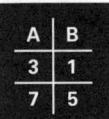 **Use or Make a Table**

 **Make an Organized List**

 **Guess and Check**

 **Use or Look for a Pattern**

 **Work Backwards**

 **Use Logical Reasoning**

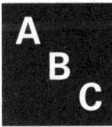 **Make It Simpler**

 **Brainstorm**

Guess and Check

1 The Bears lost the baseball game to the Lions. Use these clues to find the final score for each team.
- The sum of the scores is 18.
- Both scores are odd numbers.
- The difference between the two scores is 12.

What is the final score for each team?

FIND OUT
What do you have to find out to solve the problem?

sum is difference

What do you know about the scores of the game?

the bears lost the sum of the game the lions w... both scores are odd numbers

CHOOSE STRATEGIES
You can use **Guess and Check** to help you solve this kind of problem. Guess two numbers and then check to see if they fit the clues. If your guess is not right, make another guess. Use information from a wrong guess to help you make a better guess.

SOLVE IT

1. Begin with a guess for each team's final score.

   Guess for the Bears

   7

   Guess for the Lions

   11

2. How can you check your guesses?

   $\begin{array}{r}11\\+7\\\hline 18\end{array}$ $\begin{array}{r}11\\-7\\\hline 4\end{array}$

3. Do your guesses fit all the clues?

   no

4. Which clues do your guesses fit?

   sum of the scores are 18 both odd numbers

5. If your guesses do not fit all the clues, which ones don't they fit?

   the difference between the 2 scores is not 12

6. How can this information help you with your next guesses?

   larger number was winning and smaller was losing. to get the numbers we had to get the numbers we had to guess and check other odd number combutations

7. Keep guessing and checking until you find the final scores.

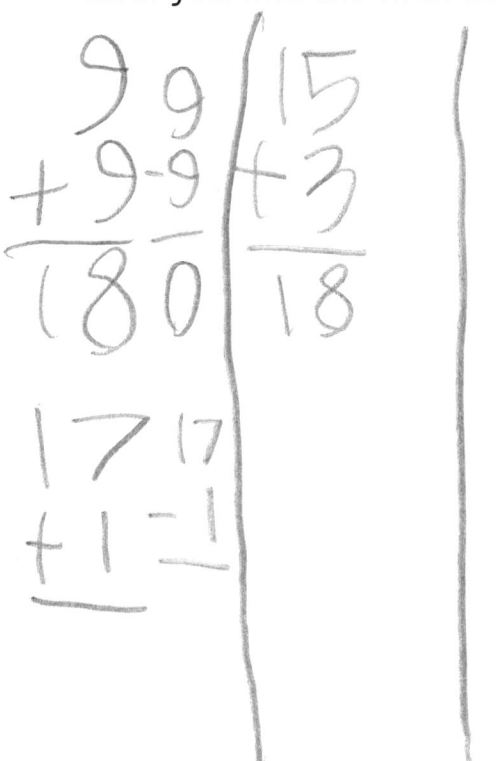

8. What is the final score for each team?

LOOK BACK
- Read the problem again.
- Check over your work.
- Did you answer the question that was asked?
- Does your answer make sense?

Guess and Check

2 These are Enrico's clues for two mystery numbers.
- The sum of the two numbers is 33.
- One number is odd and the other is even.
- One number is 2 times the other.

What are Enrico's mystery numbers?

FIND OUT
What do you have to find out to solve the problem?

What does the problem tell you about the mystery numbers?

CHOOSE STRATEGIES
You can use **Guess and Check** to help you solve this kind of problem. Guess two numbers and then check to see if they fit the clues. If your guess is wrong, make another guess. Use the information from a wrong guess to help you make a better guess.

SOLVE IT

1. Begin by guessing two numbers.

```
 13 E      13
 20 O    +13
 ──       ──
 33       26
```

2. How can you check your guesses? add them together double the small number to

3. Do your numbers fit all of the clues?

4. Which clues do your numbers fit?

5. If your numbers do not fit all the clues, which clues don't they fit?

6. How can this information help you with your next guess?

7. Keep guessing and checking until you find the mystery numbers.

```
  9    18
 +9   +9
 ──   ──
 18   27
```

```
       11
 +11  +22
 ──   ──
 22   33
```

8. What are Enrico's mystery numbers?

LOOK BACK
- Read the problem again.
- Check over your work.
- Did you answer the question that was asked?
- Does your answer make sense?

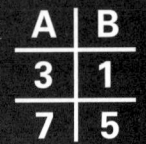

Use or Make a Table

3 Amelia's hamster is 1 month old. Her puppy is 7 months old. Some day, the puppy will be twice as old as the hamster. How old will both pets be then?

FIND OUT

What do you have to find out to solve the problem?

hamster is 1 month
puppy is 7 months

What does the problem tell you about how old Amelia's pets are now?

hamster is 1 month
puppy = 7 months

CHOOSE STRATEGIES

You can **Use or Make a Table** to help you solve this kind of problem. Make a table to keep track of the ages of the hamster and puppy.

SOLVE IT

Look at the table that has been started.

	Months									
Hamster's age	1	2	3	4	5	6	7	8	9	10
Puppy's age	7	8	9	10	11	12	13	14	15	16

1. What will you keep track of in the first row of the table?
   hamsters age

2. What will you keep track of in the second row?
   puppys age

3. Why is the word *Months* at the top of the table? because we are measuring by months

4. How old is the hamster now?
   1

5. How old is the puppy now?
   7

6. How old will the hamster be next month? 2

   Write that number in the table.

7. How old will the puppy be next month?
   8

   Is that twice as old as the hamster?
   no

8. Keep writing numbers in the table. Stop when the puppy is twice as old as the hamster.

9. How old will both pets be then?
   hamster 6
   puppy 12

LOOK BACK
- Read the problem again.
- Check over your work.
- Did you answer the question that was asked?
- Does your answer make sense?

A	B
3	1
7	5

Use or Make a Table

4 Jamie asked his sister Jacinda how old she was when she went to sports camp. Jacinda, who loved to make up puzzles, answered, "Let me think. You are 8 years old and I am 14. When I went to sports camp, I was 3 times as old as you were." How old was Jacinda when she went to sports camp?

FIND OUT

What do you have to find out to solve the problem?

how old was Jacinda was when she went to sports camp

How old are Jamie and Jacinda are now?

Jamie=8 Jacinda=14

What does the problem tell you about how old Jacinda was when she went to sports camp?

Was 3 times as old as Jamie

CHOOSE STRATEGIES

You can **Use or Make a Table** to help you solve this kind of problem. Make a table to keep track of Jamie's and Jacinda's ages.

SOLVE IT

Look at the table that has been started.

	Years					
Jamie's age	8	7	6	5	4	3
Jacinda's age	14	13	12	11	10	9

1. What will you keep track of in the first row?

    Jamie's age

2. What will you keep track of in the second row?

    Jacinda's age

3. How old is Jamie now?

    8

4. How old is Jacinda now? Write her age in the table.

    14

5. Was Jacinda older or younger than 14 when she went to sports camp? *Past tense*

    younger

6. Will you need to add years or subtract years from Jamie's and Jacinda's ages? Explain.

    Subtract because you can't go into the futre

7. How old was Jamie 1 year ago? Write that number in the table.

    7

8. How old was Jacinda 1 year ago?

    13

    Was that 3 times as old as Jamie?

    no

9. Keep writing numbers in the table. When can you stop?

    When Jacinda's age is 3 times Jamies age

10. How old was Jacinda when she went to sports camp?

    Jamie=3
    Jacinda=9

LOOK BACK
- Read the problem again.
- Check over your work.
- Did you answer the question that was asked?
- Does your answer make sense?

Act Out or Use Objects

 Play Money

5 Aisha bought a necklace at Fanny's Flea Market for 55¢. She used 9 coins to pay for the necklace. Aisha used the same number of quarters as nickels. What coins did she use?

FIND OUT

What do you have to find out to solve the problem?

She how many of eatch coin used to make 55¢ out of 9 coins

What does the problem tell you about the coins that Aisha used?

same amount of

CHOOSE STRATEGIES

You can **Act Out or Use Objects** to help you solve this kind of problem. Use play money to help you think about the problem and work out a solution.

SOLVE IT

1. What kinds of coins could Aisha have used? *quarter, dime, nicles, and pennies*

2. How many quarters and nickels could she have used? Show the coins with play money.

55¢
9 coins
(25)

3. How many more coins would she need? *7*

4. What coins could they have been? *penny and dime*

5. How many of each? Show them with play money.
5
10
25
(1)(1)(1)(1)(1) (10)(10) =25

6. Does your group of coins make 55¢?

7. Try different combinations of 9 coins until you find 9 coins that make 55¢.

8. What coins did Aisha use?
1 quarter   5 pennies
1 nickle
2 dimes

LOOK BACK
- Read the problem again.
- Check over your work.
- Did you answer the question that was asked?
- Does your answer make sense?

Act Out or Use Objects

 Play Money

6 Joel is getting ready for Crazy Hat Day at school next week. He paid 40¢ for spiders to put on his crazy hat. He used 4 coins. What coins could Joel have used?

Hint: More than one answer is possible.

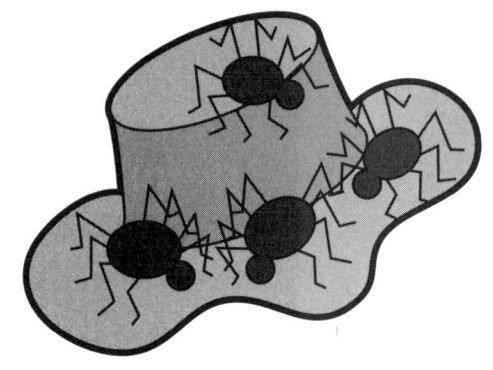

FIND OUT
What do you have to find out to solve the problem?

what 4 coins make 40¢

What does the problem tell you about the coins that Joel used? *their are 4 coins. the 4 coins make 40¢*

CHOOSE STRATEGIES
You can **Act Out or Use Objects** to help you solve this kind of problem. Use play money to help you think about the problem and work out a solution.

SOLVE IT

1. What are the different coins Joel could have used?

2. How many quarters could he have used? Show this with play money.

3. How many dimes could he have used? Show this with play money.

4. How many nickels could he have used? Show this with play money.

5. How many pennies could he have used?

6. Does your group of coins make 40¢?

7. Try putting different coins together until you find 4 coins that make 40¢.

8. What coins could Joel have used?

LOOK BACK
- Read the problem again.
- Check over your work.
- Did you answer the question that was asked?
- Does your answer make sense?

Problem Solver II 13

Make an Organized List

7 Ellie works with birds and rabbits at the Wildlife Center. On Monday she checked the legs of all the animals. Ellie looked at 14 legs in all. How many birds and how many rabbits could she have checked?

Hint: More than one answer is possible.

FIND OUT

What do you have to find out to solve the problem?

What does the problem tell you about the kinds of animals Ellie checked?

What does the problem tell you about number of legs Ellie checked?

CHOOSE STRATEGIES

You can **Make an Organized List** to help you solve this kind of problem. Use the list to keep track the number of legs as you count rabbits and birds.

SOLVE IT

Look at the organized list that has been started.

Rabbits	Legs	Birds	Legs
1	4	1	
2		2	

1. What will you keep track of on the left side of the list?

2. What will you keep track of on the right side?

3. How many legs does 1 rabbit have? How many legs will 2 rabbits have? Write that number in the list.

4. How many legs does one 1 bird have? How many legs will 2 birds have? Write those numbers in the list.

5. Keep writing numbers in the list. Stop when you have 4 rabbits and 7 birds.

6. Look in the *legs* columns. Find a number of rabbit legs and a number of bird legs that together add up to the number of legs Ellie checked.

7. How many birds and how many rabbits could Ellie have checked?

LOOK BACK
- Read the problem again.
- Check over your work.
- Did you answer the question that was asked?
- Does your answer make sense?

Make an Organized List

8 Alex and Mei counted a total of 38 legs on the spiders and grasshoppers in the science lab. Spiders have 8 legs, and grasshoppers have 6 legs. How many spiders and how many grasshoppers could there have been in the science lab?

Hint: More than one answer is possible.

FIND OUT

What do you have to find out to solve the problem?
how many legs does the

What does the problem tell you about how many legs there were? *spider=8*
grasshopper=6

What does the problem tell you about the number of legs that spiders and grasshoppers have?
spider:8
grasshopper:6

CHOOSE STRATEGIES

You can **Make an Organized List** to help you solve this kind of problem. Use the list to keep track of the number of creatures and how many legs there would be.

SOLVE IT

Look at the organized list that has been started.

Spiders	Legs	Grass-hoppers	Legs
1	8	1	6
2	16	2	12
3	24	3	18
4	32	4	24
5	40	5	30
6	48	6	36
7	56	7	42
8	64	8	48
9	72	9	54
10	80	10	60

1. What will you keep track of on the left side of the list? **spiders and spider legs**

2. What will you keep track of on the right side? **grasshoppers and grasshopper legs.**

3. How many legs does 1 spider have? How many legs will 2 spiders have? Write that number in the list.

4. How many legs does 1 grasshopper have? Write that number in the list.

5. Keep writing numbers in the list. When can you stop? **you can stop when you find a number closest to 38**

6. How many spiders and how many grasshoppers could there have been?

```
  30
+  8
____
```

LOOK BACK
- Read the problem again.
- Check over your work.
- Did you answer the question that was asked?
- Does your answer make sense?

Work Backwards
Act Out or Use Objects

 Cubes

9 Today is the Monster Marathon Race. Each team is wearing a different color. There are 4 fewer red racers than green racers. There are 4 more green racers than orange racers. There are 4 fewer yellow racers than orange. There are 2 racers wearing yellow. How many monsters are wearing each color? How many monsters are racing?

FIND OUT

What do you have to find out to solve the problem?

how many monsters are wearing each coler

What does the problem tell you about the colors that the racers are wearing?

yellow, orange, red, green

What does the problem tell you about the number of racers wearing each color?

2 yellow   4 fewer yellow than orange
4 fewer red than green   4 more green than orange

CHOOSE STRATEGIES

You can **Work Backwards** and **Act Out or Use Objects** to help you solve this kind of problem. Begin with the information given at the end of the problem. Then, work backwards to find the missing numbers. Use colored cubes to show the monsters.

SOLVE IT

1. How many racers are wearing yellow? 2

Where did you find that information? the paragraph said so.

~~Use cubes to show the yellow racers.~~

2. Now work backwards. What does the problem say about orange racers? 4 fewer yellow than orange

3. ~~How can using cubes to show the number of yellow racers help you find the number of orange racers?~~

4. How many orange racers are there? ~~Show them with cubes.~~ 6

5. Continue to work backwards to find the number of green racers. Then find the number of red racers. Use the information given in the problem to help you each time. ~~Use the cubes to show the numbers you know.~~

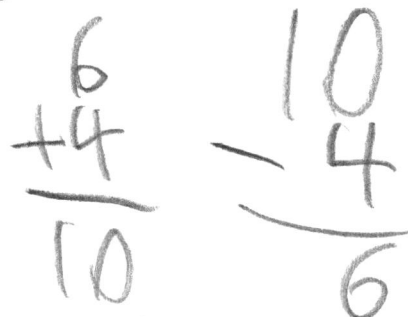

$$\begin{array}{r}6\\+4\\\hline 10\end{array} \qquad \begin{array}{r}10\\-4\\\hline 6\end{array}$$

6. How many monsters are wearing each color?
yellow = 2
orange = 6
green = 10
red =

How many monsters are racing?

2 + 6 + 6 + 10 = 24

LOOK BACK
- Read the problem again.
- Check over your work.
- Did you answer the question that was asked?
- Does your answer make sense?

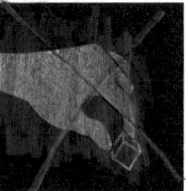

Work Backwards
Act Out or Use Objects

 Play Money

10 Jan, Martha, and Marisa wanted to share a fruit smoothie. They put their coins together to buy one. They had just enough money. They had one-half as many quarters as dimes. They had 3 more dimes than nickels, and 5 fewer nickels than pennies. They had 8 pennies. How much did the smoothie cost?

FIND OUT
What do you have to find out to solve the problem?

how mutch did the smothie cost

What does the problem tell you about the coins the girls had?

8 pennies

5 fewer nicles than pennies

one half as many quarters than dimes

3 more dimes than nicles

CHOOSE STRATEGIES
You can **Work Backwards** and **Act Out or Use Objects** to help you solve this kind of problem. Begin with the information given at the end of the problem. Then work backwards to find the missing numbers. Use play money to show the coins.

SOLVE IT

1. What kinds of coins did the girls have? **pennies, dimes, nickles, and quarters**

2. How many pennies did they have?

   Where did you find this information? **at the end**

   ~~Use your coins to show the pennies.~~

3. Work backwards. What coin is next? **nickle**

   What does the problem tell you about the nickels? **5 fewer nicles than pennies**

4. How can the information about pennies help you to find the number of nickels?

   ```
     8
   - 5
   ─────
   3 nickles
   ```

5. How many nickels were there? Show them with the coins.

6. Keep working backwards. Use your coins and what you already know to find the number of dimes and the number of quarters. **6 dimes 3 quarters**

7. How many of each kind of coin did the girls have? **8 pennies 6 dimes 3 nickles 3 quarters**

8. How much money did the girls have?

   ```
    60    15
   + 8   +75
   ─────────
    68 + 90 = 58
   ```

   Explain how you figured that out.

9. How much did the smoothie cost? **1.58**

LOOK BACK
- Read the problem again.
- Check over your work.
- Did you answer the question that was asked?
- Does your answer make sense?

**Make It Simpler
Make an Organized List**

11 Jay and his friends are having a contest. They want to see who is the best Tic-Tac-Turtle player. The 4 boys are Jay, Tom, Mark, and Alberto. Only 2 boys at a time play the game. Each boy has to play every other boy one time. How many games will the 4 boys play in all?

FIND OUT

What do you have to find out to solve the problem?

how many games they played

Who is in the contest?

Jay, tom, mark, and alberto

How many boys at a time play the game?

2

How many times does each boy play every other boy?

3

CHOOSE STRATEGIES

You can **Make It Simpler** and **Make an Organized List**. Make the problem simpler by finding out how many games just 2 boys will play. Then add one boy at a time. Make an organized list to find out how many games there will be.

22   Problem Solver II

SOLVE IT

1. Begin with just 2 boys: Jay and Tom

   Boys Playing Each Other

   Jay and Tom

2. How many games will 2 boys play?

3. Now try 3 boys: Jay, Tom, and Mark. Finish the organized list.

   Boys Playing Each Other

   Jay and Tom
   Jay and Mark
   tom and mark

4. How many games will 3 boys play? 3

5. Now try all 4 boys: Jay, Tom, Mark, and Alberto. Finish the organized list.

   Boys Playing Each Other

   Jay and Tom
   Jay and Mark
   Jay and alberto
   tom and alberto
   t and M

6. How many games will the 4 boys play in all? 6

LOOK BACK
- Read the problem again.
- Check over your work.
- Did you answer the question that was asked?
- Does your answer make sense?

**Make It Simpler
Make an Organized List**

12

Jan is swimming in the Swim Fun races. She is in a group with Hanna, Alexis, Emily, and Rita. Only 2 girls at a time will race. Each girl has to race every other girl in her group one time. How many races will there be for Jan's group?

FIND OUT
What do you have to find out to solve the problem?
how many races will there be

Who is in Jan's group?
Jan, hanna, alexis, emily, rita

How many girls at a time will race?
2

How many times does each girl race every other girl in the group?

CHOOSE STRATEGIES
You can **Make It Simpler** and **Make an Organized List**. Make the problem simpler by finding out how many races just 2 girls would have. Then add one girl at a time. Make an organized list to find out how many races there will be.

SOLVE IT

1. Begin with just 2 girls: Jan and Hanna.

Girls Racing Each Other

Jan and Hanna

J and A   h and a
J and e   h and e
J and r   h and r

2. How many races will there be for 2 girls?

3. Now try 3 girls: Jan, Hanna, and Alexis. Finish the organized list.

Girls Racing Each Other

1. Jan and Hanna   h and r / 7
2. Jan and Alexis   a and e / 8
3. J and e   a and r / 9
4. J and r
5. h and a
6. h and e

4. How many races will there be for 3 girls?

5. Now try 4 girls: Jan, Hanna, Alexis, and Emily. Finish the organized list.

Girls Racing Each Other

Jan and Hanna
Jan and Alexis
J and

6. How many races will there be for 4 girls?

7. Now try all 5 girls: Jan, Hanna, Alexis, Emily, and Rita. Finish the organized list.

Girls Racing Each Other

J and
J and
J and
J and
H and
H
H
r
r
a

8. How many races will there be for Jan's group?

LOOK BACK

- Read the problem again.
- Check over your work.
- Did you answer the question that was asked?
- Does your answer make sense?

 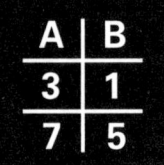

Use or Look for a Pattern
Use or Make a Table

13 When the Green Turtle Bus makes its first stop, 3 bugs get on. At the second stop, 6 bugs get on the bus. At the third stop, 9 bugs get on. At the fourth stop, 12 bugs climb on. The number of bugs that get on the bus at each stop keeps growing in the same way. How many bugs will get on at the eighth stop?

FIND OUT

What do you have to find out to solve the problem?

how many bugs will get off at the eighth stop

What does the problem tell you about the number of bugs that get on the bus at each stop?

stop 1 = 3 stop 2 = 6 stop 3 = 9 stop 4 = 12

CHOOSE STRATEGIES

You can **Use or Look for a Pattern** and **Use or Make a Table** to help you solve the problem. Look at the number of bugs getting on the bus at each stop. Use a table to keep track of the numbers. Look for a pattern in the numbers.

SOLVE IT

Look at the table that has been started.

Bugs Riding the Bus								
Stop number	1	2	3	4	5	6	7	8
Bugs getting on	3	6	9	12	15	18	21	24

multiples of 3

1. What will you keep track of in the first row?

   the stop number

2. What will you keep track of in the second row?

   bugs getting on

3. How many bugs got on the bus at the first stop?

   3

   Second stop?

   6

   Third stop?

   9

   Fourth stop?

   12

   Write those numbers in the table.

4. Is there a pattern to how the numbers change?

   x3 ✓

5. What number can you put in the table for the fifth stop?

   15

6. Keep writing numbers in the table until you get to the eighth stop.

7. How many bugs will get on the bus at the eighth stop?

   18, 21, 24

   answer = 24

   3 x8 / 24

LOOK BACK
- Read the problem again.
- Check over your work.
- Did you answer the question that was asked?
- Does your answer make sense?

 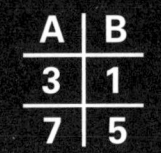

Use or Look for a Pattern
Use or Make a Table

14 Minna saved money for a Bionic Bear that cost $1.50. The first day she saved 5¢, and on the second day she saved 10¢. On the third day she saved 15¢, and on the fourth day she saved 20¢. If she kept saving in the same way, how many days did it take Minna to save $1.50?

Key = x5

FIND OUT

What do you have to find out to solve the problem?

how many days did she save up for $1.50

What does the problem tell you about the Bionic Bear?

it costs $1.50

What do you know about how Minna saved her money?

day 1 = 5   day 2 = 10   day 3 = 15   day 4 = 20

CHOOSE STRATEGIES

You can **Use or Look for a Pattern** and **Use or Make a Table** to help you solve the problem. Look at how the amount saved changes each day. Look for a pattern in the way the numbers change. Use a table to keep track of the numbers. A table also helps you find the pattern.

SOLVE IT

Look at the table that has been started.

Minna's Savings

Day	1	2	3	4	5	6	7	8	9	10	11
Amount to start	0	5¢	15¢	20¢	50¢	75¢	1.05	1.40			
Amount saved	5¢	10¢	15¢	20¢	25¢	30¢	35¢	40¢	45¢	50¢	55¢
Total savings	5¢	15¢									

1. What will you keep track of in each row?

   day
   amount to start
   amount saved
   total savings

2. How much had Minna saved at the end of Day 1?

   5

3. How much money did she start with on Day 2?

   5

4. How much did she save on Day 2?

   10

5. How much money had been saved at the end of Day 2?
   How did you find that number?

   added day 1 and the saved amount from day 2 together

6. Keep writing numbers in the table. When can you stop?

7. How many days did it take Minna to save $1.50?

LOOK BACK
- Read the problem again.
- Check over your work.
- Did you answer the question that was asked?
- Does your answer make sense?

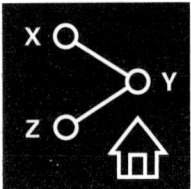

Use or Make a Picture or Diagram

15 At Family Game Night, Jade's favorite games are Funny Frogs and Catch the Cat. Last night 15 people in all played Funny Frogs and 18 people in all played Catch the Cat. Ten of the people played both games. How many people altogether played Jade's favorite games?

FIND OUT

What do you have to find out to solve the problem?

What does the problem tell you about the people who played Funny Frogs and Catch the Cat?

CHOOSE STRATEGIES

You can **Use or Make a Picture or Diagram** to help you solve this kind of problem. Use the Venn circle diagram shown.

SOLVE IT

Look at the Venn circle diagram.

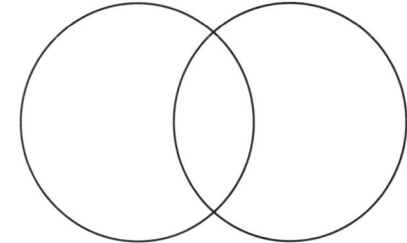

Funny Frogs   **Catch the Cat**

1. What is the label for each circle?

2. How could you label the overlapping part of the circles?

3. How many people played both games? Write that number in the overlapping part.

4. Are the 10 people who played both games part of the 15 people who played Funny Frogs?

5. How can you find the number of people who played **only** Funny Frogs?

6. How many people played **only** Funny Frogs? Write this number where it belongs.

7. Are the 10 people who played both games part of the 18 people who played Catch the Cat?

8. How many people played **only** Catch the Cat? Write this number where it belongs.

9. How can you find out how many people in all played the games?

10. How many people altogether played Jade's favorite games?

LOOK BACK
- Read the problem again.
- Check over your work.
- Did you answer the question that was asked?
- Does your answer make sense?

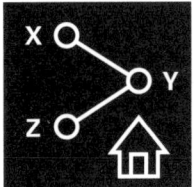

Use or Make a Picture or Diagram

16 Karla says, "I have a puzzle for you! I am thinking of 3 numbers. Together their sum is 17. One number is even. One is a number you say when you count by 3s. One number is even AND is a number you say when you count by 3s. What could the numbers be?"

Hint: More than one answer is possible.

FIND OUT

What do you have to find out to solve the problem?

What does the problem tell you about the 3 numbers?

CHOOSE STRATEGIES

You can **Use or Make a Picture or Diagram** to help you solve this kind of problem. Use the Venn circle diagram shown.

SOLVE IT

Look at the Venn circle diagram.

Even numbers   **Count by 3s**

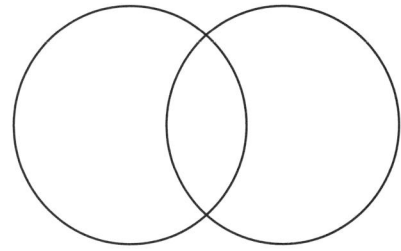

1. What is the label for each circle?

2. How could you label the overlapping part of the circles?

3. Think of a number that belongs in each part of the Venn diagram. Make sure your numbers fit the labels. Write the numbers in the diagram.

4. How can you check your answer?

5. If the numbers that you tried don't fit all of Karla's clues, try different numbers.

6. What could the numbers be?

LOOK BACK
- Read the problem again.
- Check over your work.
- Did you answer the question that was asked?
- Does your answer make sense?

**Guess and Check
Act Out or Use Objects**

 Cubes

17 Della, Tanya, and Suzi decorated 24 cupcakes for the party. Tanya decorated 3 more cupcakes than Della. Della and Suzi decorated the same number of cupcakes. How many cupcakes did each girl decorate?

FIND OUT

What do you have to find out to solve the problem?

What does the problem tell you about the number of cupcakes the girls decorated?

CHOOSE STRATEGIES

You can **Guess and Check** and **Act Out or Use Objects** to help you solve this kind of problem. Guess a number and then use your guess to figure out the other numbers. Use cubes to see if your numbers fit the clues. If your guess is not right, try again. Use the information from a wrong guess to help you make a better guess.

SOLVE IT

1. Begin by guessing a number for one of the girls. Which girl will you start with?

   What is your guess? Use cubes to show that guess.

2. How can you find the other numbers?

3. What number do you get for each girl? Use cubes to show the numbers.

4. How can you check your guesses?

5. Do your numbers fit all the clues?

6. If your numbers do not fit all the clues, what does not fit?

7. How can this information help you with your next guess?

8. Try another guess and check it. If your numbers don't fit the clues, try again. Use the information from each wrong guess to help you make a better guess.

9. How many cupcakes did each girl decorate?

LOOK BACK
- Read the problem again.
- Check over your work.
- Did you answer the question that was asked?
- Does your answer make sense?

**Guess and Check
Act Out or Use Objects**

 Cubes

18 Mark is counting the birds at the beach. He counts 39 birds in all. He sees 2 more birds on the sand than in the water. He sees 7 fewer birds in the air than on the sand. How many birds does Mark count in the water, on the sand, and in the air?

FIND OUT
What do you have to find out to solve the problem?

What does the problem tell you about the number of birds that Mark is counting?

CHOOSE STRATEGIES
You can **Guess and Check** and **Act Out or Use Objects** to help you solve this kind of problem. Guess a number and then use your guess to figure out the other numbers. Use cubes to see if your numbers fit the clues. If your guess is not right, try again. Use the information from a wrong guess to help you make a better guess.

SOLVE IT

1. Start by guessing the number of birds in one group. Which group will you start with?

   What is your guess? Use cubes to show the number.

2. Use the number from your guess to find the number of birds in each group.

3. How can you check your guesses?

4. If your numbers don't fit the clues, make a new guess. Use the information from each wrong guess to help you make better guesses. Keep guessing and checking until you find the answer.

5. How many birds does Mark count in the water, on the sand, and in the air?

LOOK BACK
- Read the problem again.
- Check over your work.
- Did you answer the question that was asked?
- Does your answer make sense?

Use Logical Reasoning
Act Out or Use Objects

 Play Money

19 The gift shop at the aquarium sells toy sea animals.

 cost the same as

If 1 sea horse costs 20¢, then how much does 1 fish cost?

FIND OUT

What do you have to find out to solve the problem?

What does the problem tell you about how much the sea horses and fish cost?

CHOOSE STRATEGIES

You can **Act Out or Use Objects** and **Use Logical Reasoning** to help you solve this kind of problem. Use play money to show what the toys cost. Use "if ... then" logical reasoning. **If** you know one thing is true, **then** you can figure out what else is true.

SOLVE IT

1. How much does 1 sea horse cost? Show that with coins.

2. How can you find out how much 3 sea horses cost?

3. How much do 3 sea horses cost? Show this with coins.

4. How much do 1 fish and 1 sea horse cost?

5. How do you know that?

6. How can you find out how much 1 fish costs?

7. How much does 1 fish cost?

LOOK BACK
- Read the problem again.
- Check over your work.
- Did you answer the question that was asked?
- Does your answer make sense?

Use Logical Reasoning
Act Out or Use Objects

 Cubes

20 In one arm, Tara is carrying 5 Mummy Mystery books. In her other arm, she is carrying her bike helmet and 2 Mummy Mystery books.

 weigh the same as

If each book weighs 3 ounces, how much does Tara's bike helmet weigh?

FIND OUT
What do you have to find out to solve the problem?

What does the problem tell you about how much the books and bike helmet weigh?

CHOOSE STRATEGIES
You can **Act Out or Use Objects** and **Use Logical Reasoning** to help you solve this kind of problem. Use cubes to show ounces. Use "if ... then" logical reasoning. **If** you know one thing is true, **then** you can figure out what else is true.

SOLVE IT

1. How much does 1 Mummy Mystery book weigh? Show that with cubes.

2. How would you use cubes to show how much 5 Mummy Mystery books weigh?

3. How much do 5 Mummy Mystery books weigh?

4. How much do 2 Mummy Mystery books and the bike helmet weigh together?

5. How could you find out how much the bike helmet weighs?

6. How much does Tara's bike helmet weigh?

LOOK BACK
- Read the problem again.
- Check over your work.
- Did you answer the question that was asked?
- Does your answer make sense?

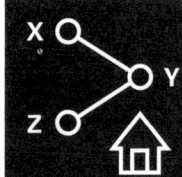

Use or Make a Picture or Diagram

21 Shari and Lamar have been told to mark these number pairs on the grid: (1, 3), (3, 5), (7, 5), (9, 3), (7, 1), (3, 1), (1, 3)

Then they will connect the points in order, drawing straight lines.

Shari says, "I think the lines will make a rectangle." Lamar says, "I think they will make a hexagon." What shape do the lines make? Who is right?

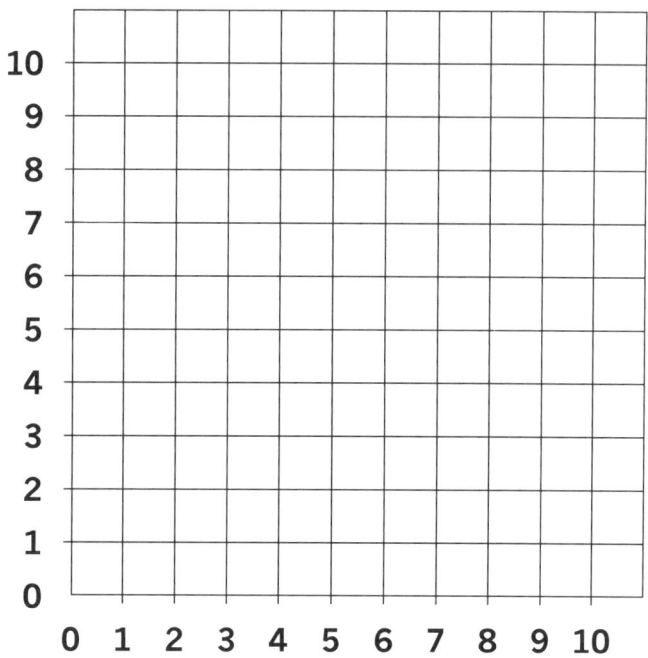

Hint: (1, 3) means 1 unit to the right of zero and 3 units up.

FIND OUT

What do you have to find out to solve the problem?

What does the hint tell you?

What do you know about connecting the points shown by number pairs?

What is the order of the number pairs given in the problem?

What shape does Shari think the lines will make?

What shape does Lamar think they will make?

CHOOSE STRATEGIES

You can **Use or Make a Picture or Diagram** to help you solve this kind of problem. Use the number pairs to find the points on the grid. Then draw lines to connect the points and make the shape.

SOLVE IT

1. What is the first number pair?

Find that point on the grid and mark it.

2. What is the second number pair?

Find that point and mark it. Now draw a straight line between the points you marked.
3. What is the third number pair?

Mark that point. Connect it to the second point.
4. Keep marking and connecting points until you make the whole shape.
5. What shape did the lines make? Who is right, Shari or Lamar?

LOOK BACK
- Read the problem again.
- Check over your work.
- Did you answer the question that was asked?
- Does your answer make sense?

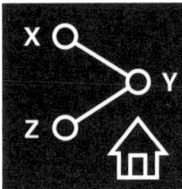

Use or Make a Picture or Diagram

22 Danger ahead! Julio must be careful or he might cross the path of the crocodile!

Use the number pairs to find each point in Julio's path. Connect the points in order. Then find each point in the crocodile's path. Connect those points in order.

Julio's path	Crocodile's path
(2, 1)	(10, 1)
(5, 4)	(8, 4)
(2, 5)	(7, 5)
(5, 7)	(9, 6)
(6, 8)	(7, 8)
(6, 9)	(6, 10)

Will Julio cross the path of the crocodile?

Hint: (2, 1) means 2 units to the right of zero and 1 unit up.

FIND OUT

What do you have to find out to solve the problem?

What number pairs are given for Julio's path?

What number pairs are given for the crocodile's path?

What does the problem say about connecting the points for each path?

What does the hint tell you?

CHOOSE STRATEGIES

You can **Use or Make a Picture or Diagram** to help you solve this kind of problem. Use the number pairs to help you place points on the map. Then draw lines to connect the points and find the paths.

SOLVE IT

1. What is the first number pair in Julio's path?

   Find this point on the grid.
2. What is the second number pair?

   Mark this point. Draw a straight line to connect this point to the first one.
3. Keep finding and connecting points this way until you come to the end of Julio's path.
4. Next, use the number pairs to find the crocodile's path.
5. Will Julio cross the path of the crocodile?

LOOK BACK

- Read the problem again.
- Check over your work.
- Did you answer the question that was asked?
- Does your answer make sense?

**Use Logical Reasoning
Act Out or Use Objects**

 | **Pattern Blocks**

23 Which of these robots is Zorbot?
Use the clues to find out.

- Zorbot has fewer than 6 angles.
- All of Zorbot's sides are the same length.
- Zorbot is a parallelogram.
- None of Zorbot's angles are right angles.

A **B** **C** **D**

FIND OUT

What do you have to find out to solve the problem?

What does the problem tell you about Zorbot?

CHOOSE STRATEGIES

You can **Act Out or Use Objects** and **Use Logical Reasoning** to help you solve this kind of problem. Use Pattern Blocks to make the robots. Use "if … then" logical reasoning. **If** you know something is true, **then** you can figure out what else is true or not true.

SOLVE IT

1. What is the first clue about Zorbot?

2. If that is true, then are there any robots that could **not** be Zorbot? Why?

3. What is the second clue?

4. If that is true, then are there any other robots that could **not** be Zorbot? Why?

5. What is the third clue?

6. If that is true, then are there any other robots that could **not** be Zorbot? Why?

7. What is the fourth clue?

8. If that is true, then is there another robot that could **not** be Zorbot? Why?

9. Which robot is Zorbot?

LOOK BACK
- Read the problem again.
- Check over your work.
- Did you answer the question that was asked?
- Does your answer make sense?

Use Logical Reasoning
Act Out or Use Objects

 | **Pattern Blocks**

24 Polly made up this puzzle.
- **I am a polygon.**
- **I have more than one line of symmetry.**
- **Trace around me, and then flip me over each of my sides. If you trace each flip, you will see a bigger me.**

Can you solve Polly's puzzle? Which shape is the one that matches all of her clues?

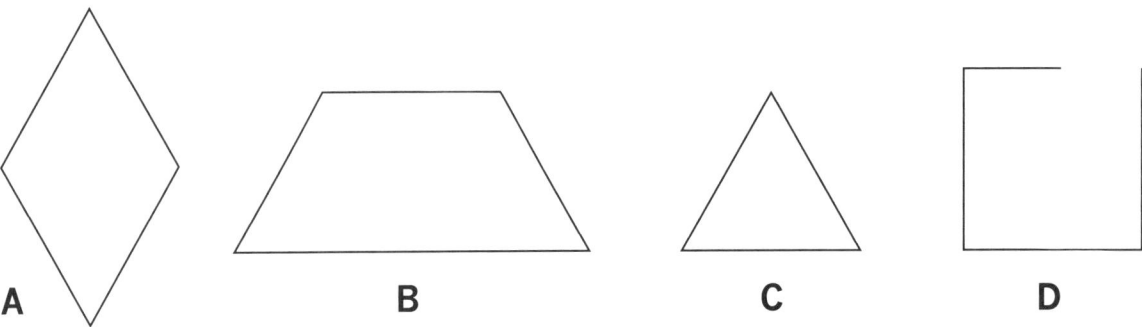

FIND OUT
What do you have to find out to solve the problem?

What does the problem tell you about the shape?

CHOOSE STRATEGIES
You can **Act Out or Use Objects** and **Use Logical Reasoning** to help you solve this kind of problem. Use Pattern Blocks to trace shapes. Use "if ... then" logical reasoning. **If** you know something is true, **then** you can figure out what else is true or not true.

SOLVE IT

1. What does Polly's first clue tell you about the shape?

2. If that is true, then are there any shapes that could **not** be the shape? Why?

3. What is the second clue?

4. How can you find out if a shape has lines of symmetry?

5. If the second clue is true, then are there any shapes that could **not** be the shape?

   Why?

6. What is the third clue?

7. Find the shape that fits the third clue.

8. Which shape fits all of Polly's clues?

LOOK BACK
- Read the problem again.
- Check over your work.
- Did you answer the question that was asked?
- Does your answer make sense?

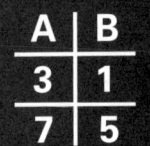  

Use or Make a Table
Use or Look for a Pattern
Act Out or Use Objects

 | Pattern Blocks

25

Jen and Kim move slowly along a hallway. They are looking for a secret door that shows the number 64. There are no numbers on the doors. On the first door, they see a small diamond. On the second door, they see a larger diamond. On the third door, they see a diamond that is even larger.

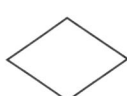 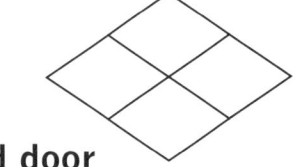

 1st door   2nd door    3rd door

"Wait!" says Jen. "I see a pattern! I know which door it will be!" Which door shows the number 64?

FIND OUT

What do you have to find out to solve the problem?

What does the problem tell you about the doors?

What do Jen and Kim see on the first three doors?

CHOOSE STRATEGIES

You can **Act Out or Use Objects, Use or Make a Table,** and **Use or Look for a Pattern** to help you solve this kind of problem. Use Pattern Blocks to build the growing diamond shapes. Make a table and look for a pattern.

SOLVE IT

Look at the table that has been started.

Door	1st	2nd	3rd		
Number of small diamonds	1				

1. How many small diamonds are on the first door?

2. How many small diamonds are on the second door? How many small diamonds are on the third door? Write these numbers in the table.

3. How many small diamonds will there be on the fourth door? Write that number in the table.

4. Look for a pattern in the numbers in your table. What is the pattern?

5. Use the pattern and keep writing numbers in the table.

6. Which door shows the number 64?

LOOK BACK
- Read the problem again.
- Check over your work.
- Did you answer the question that was asked?
- Does your answer make sense?

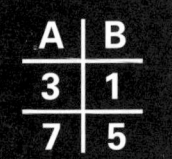  

Use or Make a Table
Use or Look for a Pattern
Act Out or Use Objects

 Cubes

26

Bug is getting ready for the big race. He races up, over, and down the first stair. "That's 3 jumps!" says Bug. He races up, over, and down the second stair. "That's 7 jumps!" says Bug. He races up, over, and down the third stair. "That's 11 jumps!" he says.

Bug keeps racing up, over, and down stairs. The stairs keep getting larger in the same way. How many jumps will he make on the eighth stair?

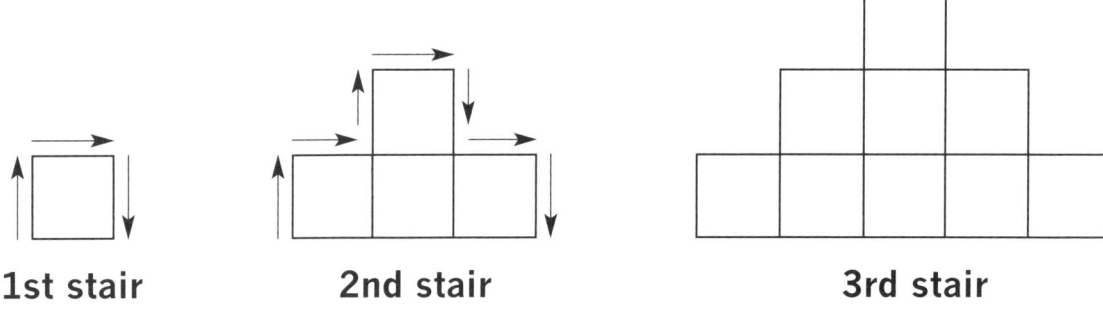

1st stair   2nd stair   3rd stair

FIND OUT

What do you have to find out to solve the problem?

What does the problem tell you about the jumps that Bug makes?

CHOOSE STRATEGIES

You can **Act Out or Use Objects, Use or Make a Table,** and **Use or Look for a Pattern** to help you solve this kind of problem. Use cubes to build the growing stairs. Make a table and look for a pattern.

52   Problem Solver II

SOLVE IT

Look at the table that has been started.

Stair	1st	2nd	3rd	4th	5th	6th	7th	8th
Number of jumps	3							

1. How many jumps does Bug make on the first stair?

2. How many jumps does Bug make on the second stair?

3. How many jumps does Bug make on the third stair?

4. How many jumps will Bug make on the fourth stair?

5. Look for a pattern in the numbers in your table. What is the pattern?

6. Use the pattern and keep writing numbers in the table.

7. How many jumps will Bug make on the eighth stair?

LOOK BACK
- Read the problem again.
- Check over your work.
- Did you answer the question that was asked?
- Does your answer make sense?

Brainstorm

27 Rosita said, "Jorge, I have a riddle for you. Why can't a rabbit, or two chickens, or you and I be in a yard at the same time?"

Jorge thought and thought. Suddenly he said, "I know!"

What was Jorge's answer to the riddle?

FIND OUT
What do you have to find out to solve the problem?

CHOOSE STRATEGIES
You can **Brainstorm** to help you solve this kind of problem. Try to think about things in different ways than you usually think about them. Try out as many different ideas as you can. The answer may pop into your mind.

SOLVE IT

1. Is there any special reason that a rabbit could not be in a yard?

2. Is there any special reason that two chickens could not be in a yard at the same time?

3. Is there any special reason that Rosita and Jorge could not be in a yard at the same time?

4. In what ways are a rabbit, two chickens, and Rosita and Jorge alike?

5. Is there more than one kind of yard?

6. What do you know about the different kinds of yards?

7. What do rabbits, chickens, people, and yards all have in common?

8. What was Jorge's answer to the riddle?

LOOK BACK
- Read the problem again.
- Check over your work.
- Did you answer the question that was asked?
- Does your answer make sense?

**Brainstorm
Act Out or Use Objects**

 | Pattern Blocks

28 **Devon gave Latanya a puzzle to solve. He used 6 square tiles to make an L-shape like the one shown. Then he asked, "How can you move just one tile and make each side of the L-shape have 4 tiles?"**

Show how Latanya can solve Devon's puzzle.

FIND OUT
What do you have to find out to solve the problem?

What does the problem tell you about Devon's puzzle for Latanya?

CHOOSE STRATEGIES
You can **Brainstorm** and **Act Out or Use Objects** to help you solve this kind of problem. Use Pattern Block squares to make Devon's L-shape. Try to think about things in different ways than you usually think about them. Try out as many different ideas as you can. The answer may pop into your mind.

SOLVE IT

1. How many tiles does Devon's puzzle shape have?

2. How many tiles are there in the long side of the L?

3. How many tiles are there in the short side of the L?

4. What are some different ways you can arrange the tiles and have 4 tiles in both sides of the L? Try as many different arrangements as you can think of.

5. How can Latanya solve Devon's puzzle?

LOOK BACK
- Read the problem again.
- Check over your work.
- Did you answer the question that was asked?
- Does your answer make sense?

 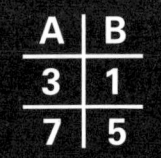

**Work Backwards
Use or Make a Table**

29

Movie	Starting Times
Ten Trolls	12:00 noon 2:30 P.M. 5:00 P.M.
Pirate Patrol	1:00 P.M. 3:30 P.M. 6:00 P.M.
RoboRats	2:00 P.M. 4:30 P.M. 7:00 P.M.

Ramal and his brother took a bus to the Movie Mall. They saw a movie that lasted 2 hours. It took them 30 minutes to get home after the movie. They got home at 6:00 P.M. Which movie did they see, and when did it start?

FIND OUT

What do you have to find out to solve the problem?

What does the problem tell you about the movie the boys saw?

How long did it take them to get home?

When did they get home?

CHOOSE STRATEGIES

You can use **Work Backwards** and **Use or Make a Table** to help you solve this kind of problem.

SOLVE IT

1. Begin with the information given at the end of the problem. What time did the boys get home?

2. How long did it take them to get home?

3. Count back 30 minutes from 6:00 P.M. What time did they start for home?

4. Work backwards again. What do you know about the movie they saw?

5. Count back that amount of time. What time is that?

6. Look at the table of starting times. Does one of the starting times match your time?

7. Which movie did they see, and when did it start?

LOOK BACK
- Read the problem again.
- Check over your work.
- Did you answer the question that was asked?
- Does your answer make sense?

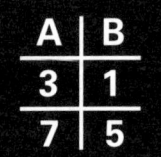

Work Backwards
Use or Make a Table

30

Camp Activities	
Swimming	10:00 – 12:00 noon 12:00 – 2:00 P.M.
Tennis	9:00 – 10:00 A.M. 10:00 – 11:00 A.M. 11:00 – 12:00 noon
Kayaking	9:00 – 11:00 A.M. 11:00 – 1:00 P.M.

Amber and Chandra are at Big Bear Summer Camp. Today, Amber did one activity that lasted 60 minutes, and then another activity for 2 hours. Chandra did one activity for 2 hours, and then another activity for 1 hour. Amber and Chandra both finished their activities at 12:00 noon. What did each girl do this morning?

FIND OUT

What do you have to find out to solve the problem?

What does the problem tell you about Amber's activities?

What do you know about Chandra's activities?

At what time were the girls finished with their activities?

Problem Solver II

CHOOSE STRATEGIES
You can **Work Backwards** and **Use or Make a Table** to help you solve this kind of problem.

SOLVE IT

1. When you work backwards, where should you start?

2. When did Amber and Chandra finish their activities?

3. Working backwards, what do you know about Chandra's second activity?

4. If you count back from noon, when did Chandra's second activity begin?

5. What do you know about Chandra's first activity?

6. Count back again. What time did her first activity begin?

7. Look at the table. What activities could Chandra have done?

8. Now go through the same steps for Amber. What activities could Amber have done?

9. What did each girl do this morning?

LOOK BACK
- Read the problem again.
- Check over your work.
- Did you answer the question that was asked?
- Does your answer make sense?

Problem Solver II   61

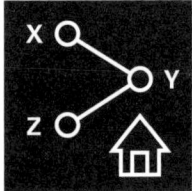

Use or Make a Picture or Diagram
Use or Look for a Pattern

31

A slug begins to climb a plant. The plant is 2 feet high. The slug climbs up 6 inches and then stops to rest. While it rests, it slides back down 2 inches. Then the slug climbs up 6 inches again and rests. While it rests, it slides down 2 inches. The slug keeps climbing and sliding in the same way. How many times will it climb before it reaches the top of the plant?

FIND OUT

What do you have to find out to solve the problem?

What does the problem tell you about the plant?

What does the problem tell you about the way the slug climbs and slides?

CHOOSE STRATEGIES

You can **Use or Make a Picture or Diagram** and **Use or Look for a Pattern** to help you solve this kind of problem. Draw a diagram to help you see where the slug is on the plant after each climb and slide. Look for a pattern in the diagram.

SOLVE IT

Look at the diagram that has been started.

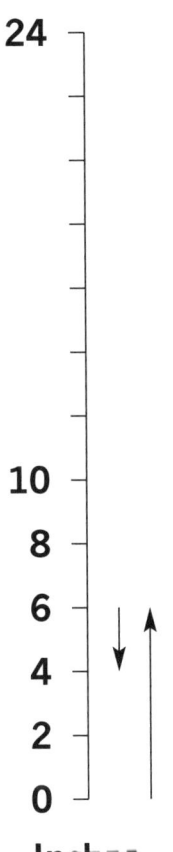

Inches

1. How high is the plant? How many inches is that?

2. Find that number at the top of the diagram. Fill in the rest of the numbers.

3. How far up does the slug go on its first climb? Find the arrow that shows this.

4. How far down does the slug go on its first slide? Find the arrow that shows this.

5. How far up does the slug go on its second climb? Draw an arrow to show this.

6. How far down does the slug go on its second slide? Draw an arrow to show this.

7. Look for a pattern. Use the pattern and finish the diagram.

8. How many times will the slug climb before it reaches the top of the plant?

LOOK BACK

- Read the problem again.
- Check over your work.
- Did you answer the question that was asked?
- Does your answer make sense?

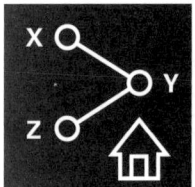

Use or Make a Picture or Diagram
Use or Look for a Pattern

32

A duck paddles slowly toward a dock that is 10 yards away. The duck paddles forward 6 feet. Then a wave pushes him back 3 feet. He paddles forward 6 feet again. Another a wave pushes him back 3 feet. He keeps paddling forward, and the waves keep pushing him back in the same way. How many times will he have to paddle forward to get to the dock?

FIND OUT

What do you have to find out to solve the problem?

How far is the duck from the dock?

What does the problem tell you about the way the duck paddles forward and gets pushed back?

CHOOSE STRATEGIES

You can **Use or Make a Picture or Diagram** and **Use or Look for a Pattern** to help you solve this kind of problem. Draw a diagram to help you see where the duck is after each paddle forward and push back. Look for a pattern in the diagram.

SOLVE IT

Look at the diagram that has been started.

Yards   0   1   2   3   4

1. How far away from the dock is the duck when he begins to paddle?

2. How far does the duck go on his first paddle forward?
   How many yards is that?
   Find the arrow that shows this.

3. How far back does the duck go on his first push back?
   How many yards is that?
   Draw an arrow to show this.

4. How far does the duck go on his second paddle forward?
   Draw an arrow to show this.

5. How far back does the duck go on his second push back?
   Draw an arrow to show this.

6. Complete the diagram. Look for a pattern that can help you.

7. How many times will the duck have to paddle forward to reach the dock?

LOOK BACK
- Read the problem again.
- Check over your work.
- Did you answer the question that was asked?
- Does your answer make sense?

 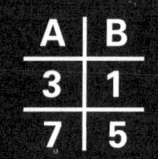

Use Logical Reasoning
Use or Make a Table

33

Amy, Ming, Carlo, and Bill went to Mona's farm. They saw horses, cows, pigs, and chickens. Each child chose a favorite farm animal. No one chose the same animal. Use the clues to match the children with their favorite animals.

- Ming's favorite animal does not have 4 legs.
- Bill's favorite animal likes to cool off by rolling in mud.
- The horse is not a boy's favorite animal.

What is each child's favorite farm animal?

FIND OUT
What do you have to find out to solve the problem?

What animals does the problem tell you the children saw?

What do the clues tell you?

CHOOSE STRATEGIES
You can **Use Logical Reasoning** and **Use or Make a Table** to help you solve this kind of problem. Use "if ... then" logical reasoning. **If** you know something is true, **then** you can figure out what else is true. Use a logic chart to keep track of your thinking. A logic chart is a kind of table.

SOLVE IT

Look at the logic chart.

	Cow	Pig	Horse	Chicken
Ming			No	Yes
Bill				No
Amy				
Carlo				

1. What will you keep track of in the rows and columns?

2. Begin with the first clue. If Ming's favorite animal does not have 4 legs, then how many legs does it have?

3. What do you think Ming's favorite animal is?

4. The **yes** in Ming's row shows her favorite animal. If her favorite animal is the chicken, then can any other animal be her favorite?

   Write **no** in all the other columns for Ming.

5. If you know the chicken is Ming's favorite, then can it be the favorite of any other children?

   To show this, write **no** in all the other children's rows under *Chicken*.

6. Look at the next clue. Write **yes** in the chart to show Bill's favorite. Then write **no** where it belongs.

7. Use the last clue. Fill in the rest of the chart.

8. What is each child's favorite farm animal?

LOOK BACK
- Read the problem again.
- Check over your work.
- Did you answer the question that was asked?
- Does your answer make sense?

Problem Solver II   67

 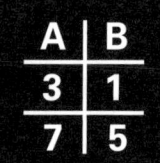

Use Logical Reasoning
Use or Make a Table

34

Midori, Wendy, Luis, Tony, and Rita are all wearing different costumes for a party. They are dressed as a ghost, a witch, a robot, a pumpkin, and a pirate. Use the clues to match the children with their costumes.

- Wendy is carrying a broom.
- Midori helps Tony put a patch over one eye.
- The first letter of Rita's name matches the first letter of her costume.
- Luis is not wearing an orange costume.

What costume is each child wearing?

FIND OUT

What do you have to find out to solve the problem?

What are the costumes the children are wearing?

What do the clues tell you?

CHOOSE STRATEGIES

You can **Use Logical Reasoning** and **Use or Make a Table** to help you solve this kind of problem. Use "If ... then" logical reasoning. **If** you know something is true, **then** you can figure out what else is true. Use a logic chart to keep track of your thinking. A logic chart is a kind of table.

68   Problem Solver II

SOLVE IT

Look at the logic chart.

	Witch	Robot	Ghost	Pumpkin	Pirate
Midori	No				
Wendy	Yes	No			
Luis					
Tony					
Rita					

1. What will you keep track of in the rows?

2. What will you keep track of in the columns?

3. Begin with the first clue. If Wendy is carrying a broom, then which costume do you think she is wearing?

   The **yes** in Wendy's row shows which costume she is wearing.

4. If Wendy is wearing the witch costume, then can she be wearing any other costume?

   Write **no** in all the other columns in Wendy's row.

5. If Wendy is wearing the witch costume, then can any other child be wearing it?

   Write **no** in all the other rows under *Witch*.

6. Continue to think about each clue and complete the chart.

7. What costume is each child wearing?

LOOK BACK
- Read the problem again.
- Check over your work.
- Did you answer the question that was asked?
- Does your answer make sense?

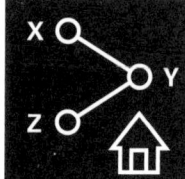

Use or Make a Picture or Diagram

35 Yolanda is very busy. She has piano lessons, piano practice, soccer practice, and swim team. Yolanda made a graph to show how many days she does each activity during one month. She forgot to label the activities! Read the clues. Then label each bar in the graph with the name of the right activity.

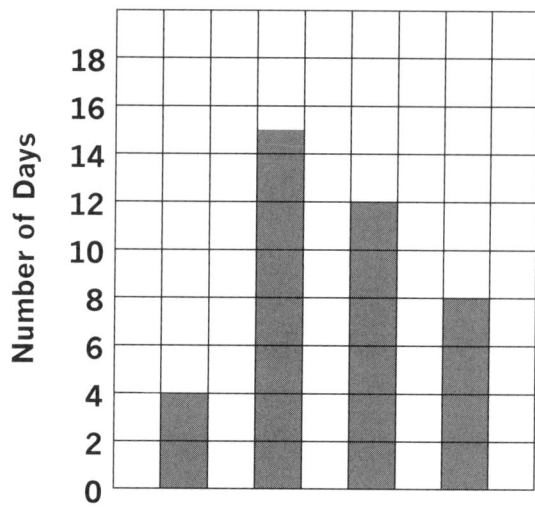

My Activities for One Month

Activities:

- She has a piano lesson every Saturday.
- She has swim team on Monday, Wednesday, and Friday.
- She has soccer practice on Tuesday and Thursday.
- She has piano practice every other day.

How many days did Yolanda do each activity during the month?

FIND OUT

What do you have to find out to solve the problem?

What does the problem tell you about Yolanda's activities?

What do the clues tell you?

70 Problem Solver II

CHOOSE STRATEGIES

You can **Use or Make a Picture or Diagram** to help you solve this kind of problem. A bar graph is a kind of diagram.

SOLVE IT

1. Look at the bar graph. What does each bar show?

2. What labels will you use for the bars in the graph?

3. What do the numbers on the side of the graph show?

4. What is the first clue?

5. How many days in a month would she do this?
   Look for a bar that matches that number and label it.

6. What is the next clue?

7. How many days in a month would she do this?
   Look for a bar that matches that number and label it.

8. Continue to think about the clues. Find and label the bar on the graph that matches each clue.

9. How many days did Yolanda do each activity?

LOOK BACK
- Read the problem again.
- Check over your work.
- Did you answer the question that was asked?
- Does your answer make sense?

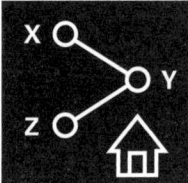

Use or Make a Picture or Diagram
Use or Look for a Pattern

36

Jimmy and Jerry are baby elephants at the Porter Zoo. They love to eat! The graph shows how many pounds of food they ate each month, starting in January. Suppose the amount they eat keeps growing in the same way. How many pounds of food do you think the elephants will eat in August? How many pounds will they eat in October?

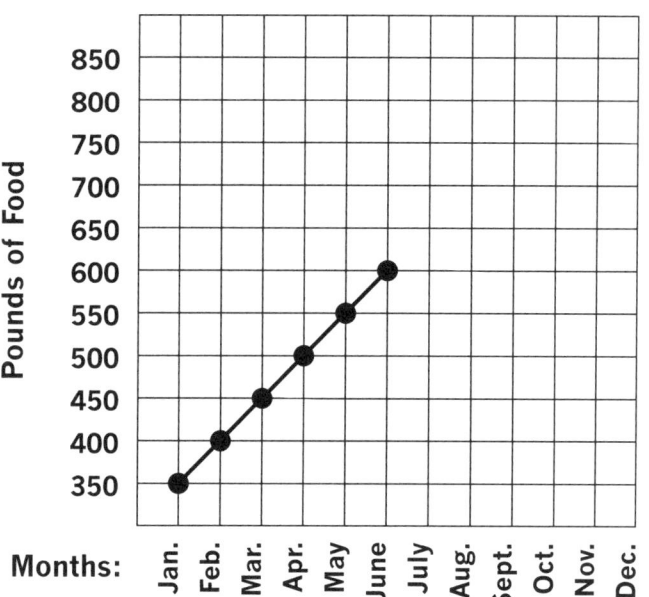

FIND OUT

What do you have to find out to solve the problem?

What does the problem tell you about Jimmy and Jerry?

What does the graph show you?

72 Problem Solver II

CHOOSE STRATEGIES
You can **Use or Make a Picture or Diagram** and **Use or Look for a Pattern** to help you solve this kind of problem. A line graph is a kind of diagram.

SOLVE IT
1. Look at the line graph. What is shown along the bottom of the graph?

2. What do the numbers along the side of the graph stand for?

3. How many pounds of food did Jimmy and Jerry eat in January?

4. How many pounds of food did they eat in February?

5. How many pounds of food did they eat in March?

6. Keep looking at the graph to see how many pounds of food they ate each month. Look for a pattern in the numbers. What is the pattern?

7. Use the pattern to help you find the answers.
8. How many pounds of food do you think the elephants will eat in August?

   How many pounds will they eat in October?

LOOK BACK
- Read the problem again.
- Check over your work.
- Did you answer the question that was asked?
- Does your answer make sense?

Problem Solver II   73

Use Logical Reasoning
Act Out or Use Objects

 Cubes

37 Ginny secretly put 6 cubes into a paper bag. Then she asked Phil to figure out the colors of the 6 cubes from these clues.

If you take 1 cube out of the bag without looking,
- the cube is more likely to be blue than red,
- the cube is least likely to be green, and
- the chance of the cube being red is 2 out of 6.

Help Phil solve Ginny's puzzle. What are the colors of the cubes in the bag? How many are there of each color?

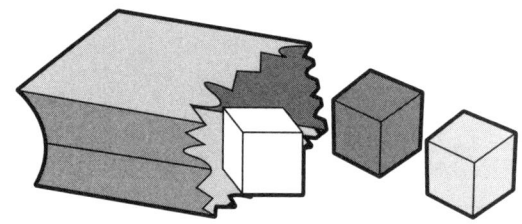

FIND OUT
What do you have to find out to solve the problem?

What does the problem tell you about the cubes in the bag?

CHOOSE STRATEGIES
You can **Act Out or Use Objects** and **Use Logical Reasoning** to help you solve this kind of problem. Use colored cubes to show the cubes in the bag. Use "if ... then" logical reasoning. **If** you know something is true, **then** you can figure out what else is true or not true.

74 Problem Solver II

SOLVE IT

1. How many cubes did Ginny put into the bag?

2. What does the first clue tell you?

3. If you know that, then what do you know about how many blue cubes there are?

4. What does the second clue tell you?

5. If you know that, then what do you know about how many green cubes there are?

6. What does the third clue tell you?

7. If you know that, then how many red cubes are in the bag?

   How many green cubes are in the bag?

   How many blue cubes are in the bag?

8. What are the colors of the cubes in the bag?

   How many cubes of each color are there?

LOOK BACK
- Read the problem again.
- Check over your work.
- Did you answer the question that was asked?
- Does your answer make sense?

Use or Make a Table
Act Out or Use Objects

Paper Clips

38 Janita and Leon are playing a game. They take turns spinning. If the spinner lands on green, Leon scores 1 point. If the spinner lands on red, Janita scores 1 point. If the spinner lands on orange, both players score 1 point. The first player to have 15 points is the winner. Is the game fair? Do the players have an equal chance of winning?

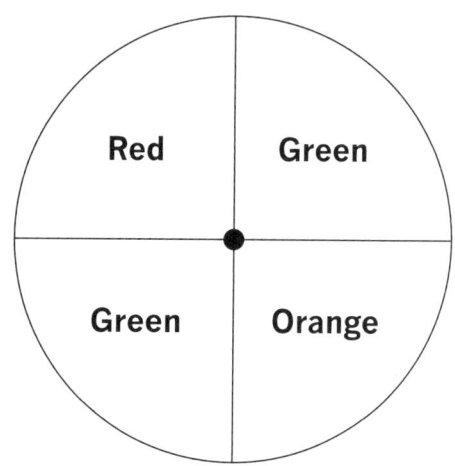

FIND OUT

What do you have to find out to solve the problem?

What does the problem tell you about how the players play the game and score points?

What do you know about the spinner they use?

CHOOSE STRATEGIES

You can **Act Out or Use Objects** and **Use or Make a Table** to help you solve this kind of problem.
Use a paper clip and pencil for spinning on the spinner.
Make a table to keep track of the players' scores.

SOLVE IT

1. What colors are on the spinner?

2. How many sections does the spinner have?

    Is each section the same size?

3. How many of the sections are red?

    What is the chance of the spinner landing on red?

4. How many of the sections are green?

    What is the chance of the spinner landing on green?

5. How many of the sections are orange?

    What is the chance of the spinner landing on orange?

6. Who do you predict will win the game?

    Why?

7. Test your prediction. Play the game two times with a partner. Let one person be Janita and the other Leon. Use a pencil and paper clip to spin on the spinner. Keep a tally of your scores in the table.

	Game 1	Game 2
Janita		
Leon		

8. Who won the two games?

9. Do the players have an equal chance of winning?

    Is the game fair?

LOOK BACK
- Read the problem again.
- Check over your work.
- Did you answer the question that was asked?
- Does your answer make sense?

Thinking Questions

Questions to think about as you are solving problems

FIND OUT
What is happening in the problem?
What do I have to find out to solve the problem?
Are there any words or ideas I don't understand?
What information can I use?
Am I missing any information that I need?

CHOOSE STRATEGIES
Have I solved a problem like this before?
What strategies helped me solve it?
Can I use the same strategy for this problem?

SOLVE IT
What information should I start with?
Do I need to add, subtract, multiply, or divide?
How can I organize the information that I use or find?
Is the strategy I chose helpful?
Would another strategy be better?
Do I need to use more than one strategy?
Is my work easy to read and understand? Is it complete?

LOOK BACK
Did I answer the question asked in the problem?
Is more than one answer possible?
Is my math correct?
Does my answer make sense? Is it reasonable?
Can I explain why I think my answer is correct?